Solid, Liquid, Gas: What Is Matter?

Erica Smith

The Rosen Publishing Group's
READING ROOM
Collection™

New York

Published in 2003 by The Rosen Publishing Group, Inc.
29 East 21st Street, New York, NY 10010

Copyright © 2003 by The Rosen Publishing Group, Inc.

First Library Edition 2003

Book Design: Haley Wilson

Photo Credits: Cover, p. 1 © Gerard Fritz/FPG International; p. 4 © Kindra Clineff/International Stock; p. 7 © Bob Firth/International Stock; p. 8 © Steve Hix/FPG International; p. 11 © Hardie Truesdale/International Stock; p. 12 © Bill Losh/FPG International; p. 13 © Mella Panzella/ Animals Animals; p. 15 © Vesey Vanderburgh/International Stock; pp. 16–17 © Fukuhara, Inc./Corbis; pp. 18–19 © VCG/FPG International; p. 21 © Jeffery Sylvester/ FPG International/Mark Bolster/International Stock; p. 22 © Faidley/Agliolo/ International Stock.

Library of Congress Cataloging-in-Publication Data

Smith, Erica.
  Solid, liquid, gas : what is matter? / Erica Smith.
     p. cm. — (Rosen Publishing Group's reading room collection)
Includes index.
Summary: Examines atoms and the states of matter, including what some scientists call the fourth state of matter, plasma.
  ISBN 0-8239-3738-0 (library binding)
  1. Matter—Properties—Juvenile literature. [1. Matter—Properties.]
I. Title. II. Series.
  QC173.36 .S65 2002
  530.4—dc21
                              2001007512

Manufactured in the United States of America

For More Information
Chem4Kids.com:Matter:States
http://www.chem4kids.com/files/matter_states.html

Contents

What Is Matter?   **5**

Solid to Liquid to Gas   **9**

Solids   **13**

Liquids   **17**

Gases   **20**

More Matter?   **22**

Glossary   **23**

Index   **24**

What Is Matter?

"Matter" is a word that describes what makes up all things on Earth. All things are made of tiny parts called **atoms**. For example, air is made of atoms of **oxygen** and **nitrogen**. Water is made of atoms of **hydrogen** and oxygen.

Matter can stay the same, but it can also change. Matter can change shape. Clay is made up of matter that changes shape. You can take a lump of clay and change its shape over and over again. A pile of sand can also change shape. You can build a sand castle on the beach and then watch the waves wash it away.

Matter can also change form. That means it can completely change its **physical** state, or the way it looks and feels. Matter can change form without changing what it is made of!

All things on Earth are made out of matter—sand, water, toys, even people.

Water is an example of matter that changes shape and form. The shape of water depends on the shape of whatever is holding it. Water can take the form of a liquid in ponds, lakes, and oceans, and when it falls from the sky as rain.

Water takes the form of a solid when it becomes ice. When the **temperature** falls below a certain level, water freezes to form ice.

Water can also take the form of a gas, or a part of the air that no one can see. When the temperature rises above a certain level, water becomes part of the air in the form of a gas. This form of water is called **water vapor**. Fog and steam are forms of water vapor that we can see. How can all of these forms be water?

Matter changes shape and form all the time.

Solid to Liquid to Gas

Water is able to take different forms because of its atoms. Its atoms act differently depending on the temperature.

Cold temperatures make the atoms in water move very slowly. The colder the temperature is, the slower the atoms in water move. When it is below 32 degrees **Fahrenheit**, the atoms in water become arranged in very neat patterns. This is when water freezes into ice. The bonds between atoms are very strong when water is frozen. That is why it is very hard to break an ice cube.

When the temperature rises above 32 degrees Fahrenheit, the bonds between the atoms break apart and ice becomes liquid water. This process is called melting. As ice melts, a solid becomes a liquid.

An iceberg is water in a solid form. An iceberg can even float in water.

You can watch things melt in nature. Sometimes snow will stay on the ground for only a short time. Snow may fall in the early morning when it is still dark outside. During the day, the sun may come up and warm the ground. This causes the snow to melt.

If the temperature gets hot enough, the water from the melted snow becomes a gas in the form of water vapor. This happens when water dries up from the ground and becomes part of the air again. This process is called **evaporation**. Boiling is another way for water to turn into water vapor. As liquid water turns into water vapor, the bonds between its atoms become even weaker, and they drift even farther apart.

Water boils at 212 degrees Fahrenheit. Boiling water will evaporate if it is not removed from the stove.

Solids

Water is not the only example of matter. Solids, liquids, and gases are everywhere you look.

Solids are the easiest examples of matter for you to feel. Sit at your desk at school. Touch your desktop, your notebook, or your pencil. These are all solids. They have a shape and a form. Different solids feel different to the touch. Some solids are hard. Some solids are soft. Some solids are smooth, while others are rough.

You can play a game with a friend. First, ask your friend to collect ten objects and place them on a plate or tray. Do not look at the objects your friend chooses. Close your eyes and ask your friend to put an object in your hand. With your eyes still closed, try to describe what you feel.

A dog's fur is soft, but a lizard's scales are rough.

15

Are the objects bumpy or smooth? Are they soft or hard? Are they big or small? What kind of shapes do they have?

After you feel each object, try to guess what it is. Have your friend write down your guesses. When you are done guessing what all the objects are, open your eyes and check your answers. How close were your guesses?

Some solids, like a baseball, are round and hard. Some solids, like this book, are flat and easy to bend. Other solids, like a pencil, are long and thin. By playing this game, you can see just how different solid objects can be.

If you closed your eyes and felt a coin, could you tell if it was a nickel or a dime? Solids feel different to the touch.

Liquids

Liquids are very different from solids. They cannot be held and touched as easily as solids can be. The form of a liquid always flows, or changes shape. A liquid is usually poured into a container, such as a cup, so that it can be held more easily.

Water is probably the liquid you know the best. Water is the most common liquid on Earth. Water falls from the clouds in the form of rain and fills our oceans, lakes, and rivers. Rain also gives us water to drink.

There are many liquids besides water. Some of them, such as **mercury**, are liquids that make up Earth. Other liquids, such as milk, are used as food. Blood is a liquid inside our bodies that keeps us alive.

You could try to hold a liquid in your hands, but it might get a little messy!

Did you know that rock can be a liquid? The temperature has to be about 2,000 degrees Fahrenheit or hotter in order for rock to become a liquid. That's nearly ten times hotter than the temperature of boiling water!

Liquid rock comes out from Earth's surface when a **volcano** erupts. Hot, flowing liquid rock called **lava** comes from deep beneath Earth's surface. Sometimes lava from deep within Earth rises and is pushed up and out of Earth. When lava cools, it forms solid rock and becomes part of the ground.

Liquid rock comes from 50 to 100 miles below Earth's surface.

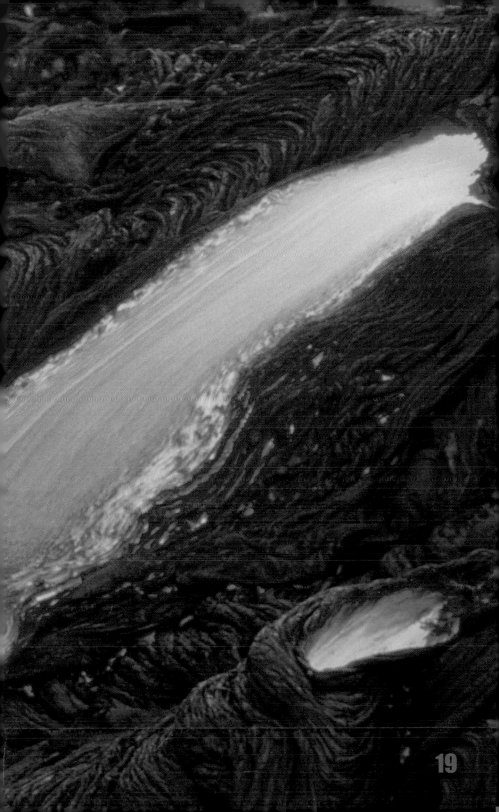

Gases

Gases are the hardest forms of matter to see and touch. Most gases cannot be seen. We cannot see the matter that makes up the air around us. The gases that surround Earth are called the **atmosphere** (AT-mus-feer).

Although the atmosphere has no shape or form, we can see its power when the wind blows. Wind can be harmful. It can also help us to move things, like sailboats and windmills.

Many gases come from nature. Plants and animals help each other by putting important gases into the air. Plants make a gas called oxygen that people need to live. We breathe in oxygen from the air. People breathe out a gas called **carbon dioxide**. Plants need carbon dioxide to make food for themselves.

We can see the force of the wind when it blows the leaves on trees. It can even knock trees over!

We can also see the force of wind when it moves the blades of windmills. Windmills are used to make electricity.

More Matter?

Some scientists say there is a fourth state of matter called plasma (PLAZ-muh). Plasmas are similar to gases, but plasmas are hotter and have extra **energy**. When a gas is given extra energy, its atoms break apart into smaller parts. A plasma is made up of these smaller parts.

You have seen plasma in the form of lightning. Stars in the night sky are another form of plasma. Their energy burns brightly from billions of miles away.

Matter is all around: solids, liquids, gases, and plasmas. Can you find examples of these states of matter in the world around you?

Glossary

atmosphere — The layer of air, gases, and dust that surrounds Earth.

atom — One of the tiny bits of matter that make up everything on Earth.

carbon dioxide — A gas that plants take in from the air and use to make food for themselves.

energy — The power to work or act.

evaporation — The process of water becoming a gas.

Fahrenheit — A scale used to measure how hot or cold something is.

hydrogen — A gas that is found in water.

lava — Hot liquid rock that comes from a volcano.

mercury — A silver metal that is a liquid at room temperature.

nitrogen — A gas that is found in air.

oxygen — A gas in air and water. People need to breathe oxygen to live.

physical — Having to do with whether something is a solid, a liquid, or a gas.

temperature — How hot or cold something is.

volcano — An opening in Earth's crust through which melted rock is sometimes forced.

water vapor — Water in the form of a gas.

Index

A
air, 5, 6, 10, 20
atmosphere, 20
atoms, 5, 9, 10, 22

C
carbon dioxide, 20

E
energy, 22
evaporation, 10

G
gas(es), 6, 10, 13, 20, 22

H
hydrogen, 5

I
ice, 6, 9

L
lava, 18
liquid(s), 6, 9, 10, 13, 17, 18, 22

N
nitrogen, 5

O
oxygen, 5, 20

P
plasma(s), 22

S
solid(s), 6, 9, 13, 14, 17, 18, 22

T
temperature(s), 6, 9, 10, 18

V
volcano, 18

W
water, 5, 6, 9, 10, 13, 17, 18
water vapor, 6, 10
wind, 20